Innovative Pest Management

The IPM Playbook

Today, as in the past, clients want to know they are getting their money's worth. Customer expectations continue to rise and tolerance for irritants, in this case a pest, or the illusion of their presence, may drive them away, for good. We all know that when a person feels "wronged" that becomes the topic of conversation for quite awhile. Our task is to prevent that, along with pressure from various consumer groups and regulatory agencies, means having to direct our attention to solving the problem in one of two ways; first finding that magic product and using a variety of items to reduce the problem, or spending the time and through a process of data collection reducing the likelihood of pests.

Pests have remarkable adaptive qualities, able to alter behavior and chemically reduce toxicity of products, or to completely adapt to different environments for survival. Continuing to rely on the newest and latest chemistry may gain an advantage for a short time, but the pest's very existence is dependent on their ability to change. They perform this adaptation easily.

A better method is longer lasting and a more integrative approach to pest reduction and elimination, one that may require some time, but pays off in huge dividends as clients needs and concerns change and their tolerance for pests decreases even more. What I speak of is IPM or as I refer to my version, Innovative Pest Management. Utilizing the collection of data (a variety of information points throughout a facility) to form a plan of action, with sustainable goals, in an economical manner. I think each account may have at its basic level the same steps of a protocol, but its adaptation by the technician is what separates the companies who truly practice innovation and those who don't.

The society we live in is becoming increasingly anti pesticide and equally anti pest, so how do we control the pest and keep the customer, by adapting as the pests do.

Integrative pest management

Integrative pest management is a process based on the acquisition and analysis of data to formulate a treatment regimen specific for an individual facility. Pest treatments in the past few decades have been chemical based, an application of material to specific targets, for pest reduction. This approach works in general, and for a short duration, dependent on a number of factors, many beyond the control of the customer or the applicator. While pest control by material has its place, the decision of when, where, how, and most important why, is often not taken into consideration. Integrative pest management seeks to take into account the whole building and the environment, monitoring the pest activity by data collection, reducing pest's attractants and harborage areas, and then developing a strategy using the information acquired in the inspection process. In the past it was considered to be cost prohibitive to employ the methods used to provide this service to facilities, but research and a merging of science with technology has brought us new materials. This, followed up with specific recommendations, based on the pests and their needs, for further enhancements to reduce the pest viability and attraction to the building greatly impact the pest presence in a building.

When companies turn to organic products or define the system of treatment based on a product choice it isn't consistent with the theme of sustainability or Green. To truly be a Green company you have to practice IPM.

Steps in Integrative Pest Management

Data Collection
Interpretation
Environmental Adjustment
Communication
Documentation

Data collection
The basis for any inspection is the gathering of data through various means, providing the professional pest management specialist with information from which to determine the likelihood of a pest infestation. This process is the key to providing a service to the client in a way that gives them the best value for their dollar. This process of data collection can be compared to the taking of a history, doing lab work, blood pressure, and pulse by a physician. We work in the world of sick structures with definite signs and symptoms of ongoing or potential problems that clients hire us to prevent or mitigate. We will look at various aspects of the data collection process, limiting ourselves to an overview of the broad but necessary steps involved, and in which every person who deals with clients in the field needs to be proficient in.
A word of caution here, this overview is not meant to be an exhaustive demonstration of inspecting and data collection, but is meant to give you the basics; the rest is up to you.

Determining areas to inspect
When arriving at the facility you will need to identify the potential areas of concern, such as lighting, landscaping and trash collection. This will give you a good idea of the level of concern the client has for sanitation and maintenance, along with the impact geography and the environment play on the structure. Make note of any water sources, mulch areas, and vegetation by the building especially near entrances. Identifying and understanding the location of lighting and break areas can help you see potential problems once you are inside. Look to see the location of children's playscapes, parking lots, wooded areas and decking or awnings. All of these whether it's a residential customer or a commercial one, can play an important part in decisions you make about control tactics.
Prior to going into the facility, check the paperwork for information pertinent to the service. It can never be stressed enough to check your look in a mirror, your appearance demonstrates to the customer your competency. Make sure you have your entire cross market materials for your client, they deserve to know the best solutions for any concerns they have and that you can supply those answers.

Gather your clipboard, service kit and any equipment you may need for the service. It is rare occurrence that once inside; anyone ever goes back to the vehicle for something they forgot. They almost always make due until next time.

Greet the customer and ask for the logbook. The customer's pest sightings and concerns should be listed in the logbook, and your attention to the customers concerns is directly proportional to their satisfaction with our service.

*A quick note on the logbook, always be sure the labels and MSDS sheets are in the logbook. If the customer has a concern about products this is where they find the info. If the logbook is missing, have another made up, it can't be stressed enough how important this communication tool is. (See appendix B)

Look for the areas pests were sighted, and any special concerns the client has. Ask about any upcoming events that may impact the protection of their facility, any changes in landscaping or construction that may be undertaken, all these could affect your service or the clients interpretation of your results.

Identify any cultural differences that may impact the treatment. We live in a very diverse cultural melting pot. Going to residences and commercial establishments puts us in front of people and their particular way of doing things. Some people may come from an area of the country where multiple families live in a 1000 SF house or friends and family members may be transient. Recognizing these characteristics and adapting your service, either by education or altering your treatment will make the client feel more comfortable and confident.

Inspection Equipment

It's very important to demonstrate your professionalism by having good equipment when you perform an inspection. The following is a short list of the basics and should be considered a minimum:

LED or Halogen Flashlight	Clipboard
Monitors/glueboards	Pen
Colored adhesive dots	Probing tool or multi tool
Magnifying device	Specimen bag or bottle

As we said, though the list looks long it's only a sample of what should be brought to an inspection. I would also look at a digital camera, knee pads, disposable tear resistant gloves, foot coverings and an extendable mirror. Carrying all of this will be a challenge and many companies use a small tool bag or belt system, either works, you just need to have the tools to do the job.

The Inspection

The inspection can often seem a big undertaking because of the volume of information you are trying to gather. The best method is to systematically divide the area you wish to inspect into zones or sections. Search top to bottom or bottom to top, moving about 5 feet at a time. If you have nothing in the section, such as a blank wall you can move pretty rapidly. Remember you are looking for moisture, food, shelter and often a lack of air movement. Air movement can sometimes cause a pest to avoid the space because of their natural desire to use air movement as a defense alarm. Continue to inspect in a pattern until the room is done. Stay in the room until it is complete, otherwise following a wall will move you to another area and often causes inspectors to miss an important item. An average room should take about 2 minutes to check, depending on the size, the amount of items in the room, the pest looking for and the customers questions. Perform an overview inspection first, one that covers everything, but allows you to focus on specific items as you move along. This method is considered prioritization of data input. Big name for an easy concept, when an EMT looks at a victim, they utilize the same principle, looking at life threatening injuries first, ABC's, and then moving to other injuries. As we have used the analogy of treating a sick building, we'll continue in that theme.

Take plenty of notes as you go, even take pictures as necessary, many computer programs will allow you to load those into the account. I often carry a small note pad to write information quickly and clean it up for the service completion form or ticket I'll leave with the customer.

Remember to look and smell because some problems may not show themselves, but the decay process can be smelled easily.

Areas you will always inspect

Kitchens	Living Rooms
Pantry	Basement
Interior entrances	Food areas
Bathrooms	Trash collection areas
Receiving areas	Locker rooms
Exterior entrance areas	Break areas
Garage	Decking
Doorways	Employee personal items
Windows	Raw product

What to look for

Evidence of pests

Live pests	Dead pests
Parts of pests	Gnaw marks or destruction

Pest debris such as sand, dirt, or nest materials

Pest Biological Needs

Moisture areas, not just water but condensation can support pests
Shelter such as empty boxes, booths, and materials in storage
Food sources, this can include human food, pet food, or even droppings

Monitors

Monitors are the 24/7 workhorses of the IPM system. Placed in the interception zones of pest paths, they are extremely effective at harvesting the occasional pest and also helping determine where the infestation is. Placement is critical in getting good performance, because the monitors can show negative results if placed in a position that is easy for the inspector to find, but not along a path preferred by the pest. Remember pests like to move along a line or wall; they like to have the feeling of security or be able to orient themselves over distances. Sometimes they just don't know any better. Always place the traps:
- in areas adjacent to cracks and crevices
- openings to void spaces
- along edges of a structure that is hidden from view
- along edges of a structure next to moisture sources
- at openings for pipes
- wherever you think its best

Check the traps regularly and use more than you think you may need. Not having enough traps can be fatal to an IPM program, it's from the data collected here, during your inspection and surveys that you put together your plan.

When reading the monitors; follow these steps:
1. Identify the pests
2. Identify the life cycle stage and sex
3. Identify the location on the trap

You will be able to do this faster as you gain experience, but it is important because this step separates the IPM practitioner and the wannabe. For example when you have adult male roaches and non gravid females on a glue board it means you are close to the main harborage, the gravid females tend to stay home. You need to investigate further or the roach problem can get out of hand quickly, if you miss this indicator of the location of the harborage. Be sure to replace the traps often also since this is a sign of

professionalism and goes a long way

ILT Monitors

Maybe not the typical way to begin, when talking about flies, but recently we have received a number of calls dealing with just that subject. It seems there is confusion among the masses about the size or area an insect light trap will cover. This is a common misconception among people looking to get value and efficiency from their purchase. We compare products all the time, how much mpg does this car get compared to that, what is the difference in quality between this product or that. This is perfectly normal and expected especially in today's marketplace where we want the most bang for the buck, however when comparing any products, services, or investments, there are many factors to consider beyond simple measurements.
Let's get back to our original question, how much area does an insect light trap cover? On the surface it should be a relatively easy and quick comparison value for rating fly lights, but as with the health care bill, there is much below the surface. Distance covered is a question regarding the pulling power or attraction of a device as it compares to another. I will look at a few characteristics here which show how this measurement can lead to a purchase which may not give you the desired results.
70% of a fly's brain is used to process visual input, with attraction to ultra-violet light being a prime stimulus. Manufacturers usually divide insect lights into 2 categories, front of the house and back of the house (For purposes of this article we will be focusing on glue-board based systems).

Ultra-violet light used with a front of the house system is usually hidden, either reflected off a substrate the light is hanging on or having some shield to prevent the general population from seeing in the fly light. We wouldn't want customers at the establishment eating and looking at flies in the dining area. Lights that reflect or to be completely correct refract the light off the wall are dealing with a matte finish or whatever the light is hanging on, which will absorb a certain amount of UV light. The less glossy the surface the more UV light is absorbed. In other words a flat paint will absorb more UV light than a gloss painted or mirrored wall. So our first consideration in choosing a front of the house system is what it will be mounted on. Refraction type systems are generally less efficient for this reason. This leaves us with a shielded system for the front of the house, as a much more efficient way of monitoring the flying insect population. Shielded systems with have a reflective area to allow the UV light to disperse more evenly away from the wall and be more attractive. The back of the house systems aren't hampered by this problem and will have exposed lamps and provide a

greater attractive surface area.

A second characteristic which must be discussed is the lamps themselves. Studies show flies are most attracted to UV in the 340 – 380 nanometer range, consistency of the operating UV is the key here. A UV light that produces a proper wavelength of UV, but fails to keep within that range for long periods of time is only effective temporarily. By temporarily we mean less than 6 months, even though manufacturers will argue length of efficacy, studies and empirical real world data shows they rarely last as long. This is a function of several characteristics inherent in the manufacture of the lamp itself. The glass used to house the lamps can be made of Woods glass, which has a mixture of lead involved in the process of manufacture and decreases the UV life, as opposed to a lamp with no lead in its makeup. Proper coating of the lamp plays a crucial role in producing a visual UV spectrum; silicate materials lose their efficiency quicker than those lamps using a fluoro-borate compound. What this means is a lamp with a non lead glass and a fluoro-borate material coating the inside will outperform and hold the UV in the proper UV wavelength longer and be more efficient.

A third consideration would be the environment the light is placed, and the competing light sources available to the insect. We have all seen how a candle burning in the dark can be seen from a great distance. Military snipers can pick out someone lighting a cigarette from ¼ mile away, giving more credence to the fact that smoking can kill. Testing done by manufacturers to give the distance covered by lights often use the best case scenario, a dark room such as a warehouse and a single light used for testing. Other overlooked factors to consider are will the light be left on for the recommended 24/7 or will it be turned off in a mistaken attempt to conserve its energy. Turning the light on and off uses up the starter material contained in the lamp faster and results in the blackening associated with worn out lamps. Will the proper maintenance be performed on a regular basis to keep the lamp running at its optimum ability and the glue-board changed as recommended to actually be catching the insects.

Survey

An often underestimated form of data collection is the employee interview. This is the opportunity to question those people who may have seen the pest and can describe things about the encounter that wouldn't fit on the pest sighting log. This isn't meant to be an interrogation but a fact finding and discussion opportunity. When looking for someone to discuss seeing pests, most people will go to the manager; this usually will result in limited or second hand info at best. The best people in the facility to question are the ones who are actually involved in the jobs which place them in contact with the pests, i.e. bus boys, dishwashers, janitors, etc. The key questions you need to ask are:

Who saw the pest?
Followed by
What did you see?
Where did you see it?
When did you see it?
Where did it go?

The importance of asking open ended questions can't be over stated since this will get the client talking and may elicit information no one thought to ask. Be sure to document all the particulars in case you need to follow up.

Interpretation of the data

The interpretation of the collected data is what separates the best IPM companies from the rest. Taking all the varied pieces of data and analyzing them is analogous to a doctor providing a diagnosis, a detective piecing clues together or a stock trader deciding which company to invest in and all at the same time. Knowledge and experience go a long way to helping with the interpretation, but it's a process that can be learned easily and quickly.

What do the pests need to survive? Food, Water, Shelter and they prefer areas with little air movement. During your data collection where did you find these areas?

Food

Food is a necessity for pests, but they can and often do get their food in a number of ways. We are back to the question of pest identification and biology. You have to know the enemy to know what they eat. Once you have identified the pest and you know the food it needs:

- Were these long term problems or singular occurrences?
- What about seasonal variation?
- Do they cook on the premises?
- Is the food bagged?
- Sanitation issues?
- Trash collection (Do they use self closing garbage cans?) How, when?
- Is there organic debris build up?

Water
Pests need water. They may be able to find nutrition from odd sources or not eat for some time, but they need moisture for survival, especially in a dry environment. Locate the areas, then ask:
- Continual moisture issues?
- A/C issues?
- Landscaping and mulching material?
- Ponds, fountains, lakes?
- Roof problems
- Leaking doors, windows or structural defects

Shelter
Shelter is also a necessary component for pest infestation. The shelter may provide food in the form of fungus or debris; sometimes other pests may become the food source. When you look at the storage ask:
- How cluttered?
- Product rotation, first in and first out?
- Housekeeping issues?
- Maintenance issues?
- Construction issues?
- Lighting outside; is it located on the building or shining onto it?

You have all this information gathered as part of a primary survey. During your data collection; you have documented any discrepancies or concerns and now have a clear picture of potential problems.
The interpretation of collected data as you see is a process, but as we said one that can be learned. The key to all good interpretation is to Keep It Simple.

How can you break the triangle?

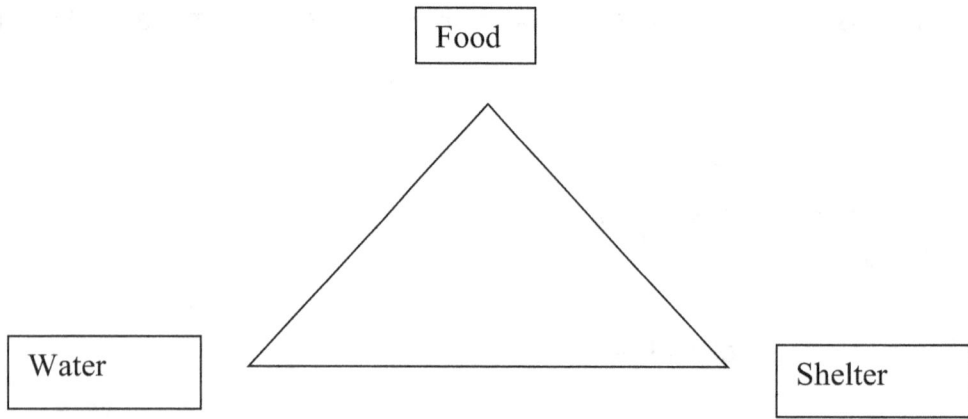

You have data on food, water and shelter. You know how each of these plays a role in the pest entrance and harborage. Now how do you put together your plan of action? How do you use the information to break the triangle? In your program you have to control these points on the triangle and the better you do it without the use of pesticides the more valuable you are to the customer. The first step in control should be the Environmental Adjustment.

Environmental Adjustment

At this stage you take the information about the structure, combine it with the pest signs and your knowledge about the biology of the pests to complete your treatment recipe. Treatment strategies depend a great deal on the company's philosophy of material use. The tide of public opinion has come full circle to a more natural control method and strategy. This doesn't mean to eliminate pesticides, but it does mean to justify and document the reasons for their use, utilizing a judicious approach to application.

Consideration must be given first to the reduction of any pests present, second to eliminating the potential for more pests to enter and lastly correct facility issues. These facility issues may entail not just structural modifications but cultural modifications via education to help in the pest elimination process.

Getting rid of the pest may involve just vacuuming or placing out snap traps and glueboards. Think first of non chemical means of getting rid of the pest, especially if they are adults.

Eliminating the pest entrance and correcting facility issues will fall under the heading of habitat modification. A few examples of this are screens on windows, sealing cracks and openings, and placing copper mesh in weep holes. Some questions to ask regarding the facility are:

- What activities can the client do to reduce pest attraction to the building or space?
- What modifications can be made to the structure to eliminate pest entrance?
- What surrounding areas can be modified to keep pests from building up in large numbers?

The environmental adjustment of the structure and its surrounding area is simply anticipating the pest's activity and taking measures to stop it. Eliminating attraction to a space or structure is easier and more effective than eliminating the source of pests. The attraction reduction anticipates the pest pressure and source reduction takes care of the pest after it has entered.

Sanitation is a key component in environmental adjustment. It is the key to the control of flies, roaches, ants and rodents, just to name a few. Customers often fail to adhere to our recommendations or to even clean, expecting you as the pest management professional to "just treat". An effective method to reduce the sanitation issue is to use Bio remediation materials to enhance the sanitation process, reducing the need for arbitrary and often unnecessary treatment with pesticides. Customers see the thoroughness of the application and feel the value of your service. Traditional pest products increase in their efficacy when no competing food sources are available. This Bio remediation treatment can be applied to cove areas, trash receptacles, drains, equipment bases, walls and any areas that won't contact food and have a deposit of fats, oils or grease (FOG). Bio remediation products are bacteria and sometimes enzymes that digest the FOG and produce water and CO_2 as waste. When they are in the presence of their "food" they double in amount of bacteria digesters every 20 minutes or so, meaning 1,000,000 bacteria become 8,000,000 in an hour. Give these hard working bacteria 3 hours and we have well over 500,000,000 bacteria digesting the FOG. That will get clean pretty quick. But it isn't an instant fix, since there are other factors that influence the growth of the bacteria and the reduction of the FOG.

 This is a great start in the reduction of food and breaks one side of the triangle.

The exterior reduction of attractants and environmental adjustment involves sanitation of the trash receptacles including dumpsters, changing lighting from mercury vapor to sodium vapor, brushing webs from overhangs and areas around entrances, and placing rodent stations with snap traps in them next to entry doors, keeping vegetation 18" from the building and improving the landscaping to pull moisture away from the building to name a few.

Once you have improved the sanitation and reduced the attractants around the building any caulking, screening, and other entrance reduction methods will provide further help.

Communication

Where inspection is the key to great pest management programs, communication is the key to keeping customers satisfied. Listening to the customer, and addressing their concerns; both those they know about and those they don't, will keep the customer happy and finding value in your service. Effective communication begins with the logbook or pest sighting log; it is the way for the client to express their concerns in writing and for the pest management professional to document their treatment of those issues. The pest sighting log must be kept in an area where both parties, client and the pest control manager, have access to it. In large facilities you might have multiple pest sighting logs.

The contents of the logbook:
Company information
Labels/MSDS
Integrative pest management overview
Treatment specs
Pest pictures
Pest sighting log
Service reports
Cross marketing materials

The maintenance of the logbook is the responsibility of the technician. How this is accomplished demonstrates to the customer your professionalism.

Service reports should detail all the relevant information about the treatment performed on that day. It is a legal document. The complete listing of materials used and recommendations for the facility must be written on the report. Remember the job isn't done until the paperwork is complete.

Lastly, you want to make sure you go over the report with someone who can get the corrections done you recommended. Just getting a signature for proof of service won't help the client. Telling the customer they need to clean won't work either, the report must be reviewed and be thoroughly gone over explaining what, when, how things need to be corrected. You have to assume if they don't understand, they won't get it done.

Selling the innovative Program

Fundamentals

Before you can play the game you need to understand the fundamentals. A universal truth for anything you undertake. Yet, why do so many people go into "sales" with no clear conception of what is necessary to be successful. Sales is not easy money, it's hard work, maybe not physically demanding, but mentally and emotionally challenging.

All of us sell everyday. Another one of those truths we don't give much thought to. In any job, except for a very few, we sell something. Maybe it's selling an idea, selling a request or the actual process of making a living at sales, we all sell. It's an inherent ability, just look at any child who wants something, first may come the request, then the demand, then the bargaining, the tantrum and sometimes with some really practiced kids, the psychological manipulation "I didn't really want it". Kids are amazing in their skills of negotiation.

At some point we lose that ability to ask for what we want. We over think, we become self-conscious, and we get scared of rejection. All of it we create in our own minds. If we put it there we can change it, it just takes practice and a change in outlook.

With that in mind we begin to look at what is involved in this career choice of selling.

All sales must begin with a basic knowledge of what you are selling. That may seem pretty straight forward, but it's amazing how many people think that once they get out in the real world, they can convince people to buy their product or service or whatever. Nothing can be further from the truth, although some people get lucky, you have to be able to explain what your company offers and more importantly how your potential customer's needs could be filled by your company.

What do you offer?
It is said that customers only listen to radio station WII FM, What's in it for me. Most do, even if they may not be aware of it, but we have to be able to help them see what is in it for them. Our job is to demonstrate that we can take care of a need or want they may not even be aware of. The first step in that process is to be aware of what you offer. You need to be able to answer these questions:

What does my company or organization produce?

What type of company or organization are we?

What do you produce? It may not be a tangible product, it may be a service, but you still produce something. It may be peace of mind, it may be reduced liability, it may be protection and enhancement of your customers property, no matter what you produce something.

What type of organization are we? Are you a service organization like a restaurant that empowers its people to make spot adjustments or give away free meals? Are you a non profit organization? Are you a manufacturer? You may be an IT/ data organization? A service company may have different lines or products it sells and each must have their own identity. What ever it may be whether producing widgets or managing knowledge, you must define your existence. This may be done in a mission statement or vision statement. (See Appendix E: Analyzing your business)

Once those questions are answered you are ready to look at the basics, the fundamentals. Everything you do as an expert, and that's what you need to be in order to sell, revolves around your organizations core beliefs and reason for being. All the cold call explanation, the closing techniques and the documentation training, come from your ability to promote your organization and its beliefs.

Blocking and Tackling
Once you know what you produce and what type of organization you are, how do you get the customers? These are the basics of selling. First of all where do you find the potential customers? Where do you go to get leads? Very often sales people believe that the opportunities to sell will magically appear, even experienced salespeople often ask where and how do I get my leads? No matter

what industry you are in, eventually the leads dry up and you end up in a desert without water, now what?
Who can best use your organization?
What are the obvious businesses or customers that everyone in your industry goes to for business?

Make a short list of these.

Where do you need accounts at to fill geographic needs in your organization?

Do you need local areas or distant satellite cities that have to be expanded?

What vendors do business with the businesses you are trying to get into? Form a list of areas and product support or raw materials used by the common businesses everyone else goes after. Now you have a starting point. You can place a target on the potential sales and visualize helping them. Where do you look for other contacts?

Old customer lists	Social Networking sites
Cancelled account list	Chamber of Commerce
Magazines	Associations
Lead lists	Internet

The list could go on, but that gives you an idea. Put your list together in a conspicuous place and don't be afraid if it contains some large or big dream type of accounts. As they say "Dream big or go home".

Preparation
What next?
How do you approach them? Make sure you have prepared the following before you venture out:

Action oriented message. This refers to spending time to develop a clear and memorable message. It involves being very clear about the benefits of what you bring to the table and "speaking the language" of your potential customers. Sales people often resort to phrases like, "I help my clients be bug free." "I do IPM." "I have a Green service" Is there anything memorable about any of these statements? Try these instead: "I help my clients protect their health and property." Or "My clients reduce their need for pesticides in and around their home" Do you see how each of these statements provides more clarity and gives the potential customer more information about benefits?

Elite mindset. Your sales presentation is most effective when you have passion and absolutely know that it's effective. You really want people to experience your service because you know it's the best and any other choice is a mistake. Some people call it an attitude and that is exactly what it is, the attitude that you have the best and you're proud of what your team can accomplish.

Be Real. You have to walk the walk and mirror your customers. People like to spend time with and have a greater trust in people they feel fit into their group.

For example, if you are trying sell your service or product to the Regional Manager of a major company, you have to look like you belong there. They must believe that your service or product is used by people at their level. They may want to be the first on the block with a new product or given a strategic advantage but they have to trust you to do it. Otherwise they are always the guy who did something incredibly dumb. No one wants to be that guy.

You have gone through a client list, made up customers you want to target and prioritized the big ones. You have set them up in tiers to devote the time as necessary to get the customer into your company. You have analyzed your company and feel strong and confident in what you offer. You have the info and product offerings and are ready to go out to make sales.

Now what?
There are a few things that all sales coaches and leaders recommend:
Research- You have to use the social; media, the newspaper, the internet and anything else to gather info about the company.
You also need to do the same for the industry that company is in. You want to become a consultant not a sales person.
If you can recite your product list you're a human catalogue,
If you can tell the customer what the products do you're a technician, If you can tell the customer how the products will help them you're a salesman and
if you can describe what they need how you can help and in what ways your products can help them you're a consultant.
That is a worthy goal to achieve, but if you can do all that and tell the customer what trends are occurring in their industry you're an expert. That's where we want to be. We want to be a part of the customer's family at work.

Dry runs- Practice what to say and what questions the customer may ask. Have your answers ready, using demonstration, stories, graphs, charts whatever gets the customers various senses engaged in the proposal. It's important to not have a canned presentation, but be ready for the most common questions.

Why should the potential client buy from you?
What makes you the provider of choice?
Do they have a need?
Are you the company to fulfill the need?

Plan- Plan what you want to accomplish at the meeting. It is important to have a goal each time you meet. This helps tell you if you have momentum or if you have stalled.

Out on your own
In service sales you have 2 stages you have to work through, first the inspection and then the proposal. The inspection in many ways may be the more difficult because of the potential clients reluctance to want a second proposal or may perceive they don't have a problem.

Here is where your attitude must come through, they need you, otherwise they are using an inferior service. Not everyone will realize this at first but keep trying, remember you're the expert. The single biggest hurdle and the most common according to sales surveys is not asking. Whether it's the inspection opportunity or during a presentation, more business is missed because of a reluctance to ask for it.

People will trust you and believe you, but you have to be able to deliver the goods. You can have the attitude, the best marketing or opening line, but if you fail to deliver you'll be dead in the water. Find out what resources you have and how to use them. To continue being extremely skilled, upgrade your education and training whenever possible. Publish your certifications and training on a social network site to add more power to your presence. Don't forget, your potential customer looks you up also.

Be sure to always actively listen, sometimes casual conversation and meetings that don't result in a sale may lead you to another opportunity. Be aware of your "silent" language. Do you look interested? Are you listening? Do you ask good questions? Do you provide the prospect an opportunity to ask questions or get more information? Do you have resources to recommend to this client if it turns out she or he is not a match for your business? Practice this, it's not as easy as it seems.

Always be consistent. Commit fully to daily marketing actions and then take them. Have a plan and work the plan, it may be easy to put something off or skip a day, but you must always be thinking of how you can provide the excellence to your customers.

A note here on time management. You always want to be making some progress, setting the next meeting date, gathering more info, doing a proposal, when this momentum stops or slows dramatically; you have a decision to make. Fish or cut bait. You can't have the potential customer become a drain on your time, there is another customer out there waiting.

Does the customer have the resources ($) for the project? They may want the project, they may have a need, but maybe they are in a merger or money is tight, be ready for that situation. Often the potential customer may use money as an excuse, but sometimes they actually have no money. At this stage you also want to keep in touch, but you have other customers that need your services. Maybe you have a creative way to help them afford your project, that's great, but don't sacrifice your long term goal of being their expert for the quick dollar of today.

This may take several sales calls because remember we are trying to find where we fit in their organization. Obviously you don't want to spend all your time at this one customer, you prioritized the accounts and are setting up meetings and follow up as you move through the sales process here.

The inspection
Data collection
The basis for any inspection is the gathering of data through various means, providing the professional pest management specialist or inspector with information from which to determine the likelihood of a possible pest infestation. This process is the key to providing a service to the client in a way that gives them the best value for their dollar. This process of data collection can be compared to the taking of a history, doing lab work, blood pressure, and pulse by a physician. We work in the world of sick structures with definite signs and symptoms of ongoing or potential problems, the client pays us to prevent or

mitigate. We will look at various aspects of the data collection process, limiting ourselves to an overview of the broad but necessary steps involved, and in which every person who deals with clients in the field needs to be proficient in.

Sales people or inspectors often do a point in time inspection, but as you can see, we recommend an initial inspection and a follow up. The initial inspection may be easier to sell the potential customer on if all you are doing is placing out some monitor devices. Your follow up time can be reduced now, since you have a location and samples to show the potential client. Many companies already use the monitor devices, which means your inspection should be easier.

Determining areas to inspect

When arriving at the facility you will need to identify the potential areas of concern, such as lighting, landscaping and trash collection. This will give you a good idea of the level of concern the client has for sanitation and maintenance, along with the impact geography and the environment play on the structure. Make note of any water sources, mulch areas, and vegetation by the building especially near entrances. Understanding the impact of lighting and smoke or break areas can aid you in locating potential problems once you are inside.

Prior to going into the facility, check the paperwork for information pertinent to the inspection. It can never be stressed enough to check your look in a mirror, your appearance demonstrates to the customer your competency. Make sure you have your entire cross market materials for your client, they deserve to know the best solutions for any concerns they have and that you can supply those answers. Gather your clipboard, inspection kit and any equipment you may need for the service.

Greet the customer and ask for the logbook. The customer's pest sightings and concerns should be listed in a logbook.

Look for the areas pests were sighted, and any special concerns the client has. Ask about any upcoming events that may impact the protection of their facility, any changes in landscaping or construction that may be undertaken, all these could affect your future service or the clients interpretation of your results.

Areas you will always inspect

Interior entrances	Food areas
Bathrooms	Trash collection areas
Receiving areas	Locker rooms
Exterior entrance areas	Break areas

What to look for

Evidence of pests
Live pests   Dead pests
Parts of pests   Gnaw marks or destruction
Pest debris such as sand, dirt, or nest materials

Pest Biological Needs
Moisture areas, not just water but condensation can support pests
Shelter such as empty boxes, booths, and materials in storage
Food sources, this can include human food, pet food, or even droppings

Potential entrance areas
Doorways
Employee personal items
Windows
Raw product

Document everything and if allowed take pictures. Many computer programs will allow you to attach the images to the account and maybe incorporate them into the proposal.

Graph

An often overlooked step and one that will help your credibility and professional look is the graph. Charting where and what you saw helps you, the technician and the customer and eliminates any questions about what they need or what you provide. It may not be an architectural masterpiece, but a good graph is an important tool.

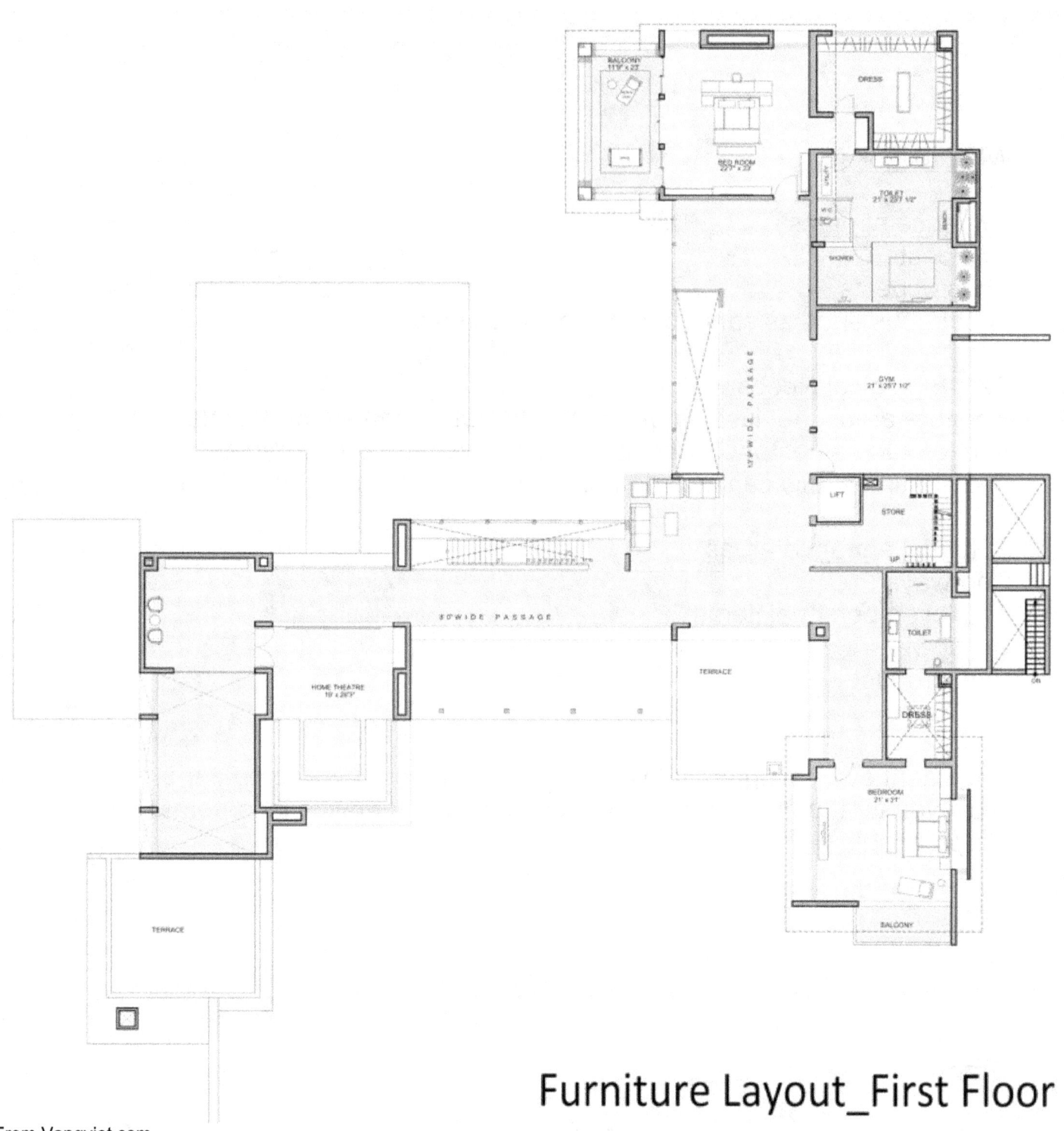

Furniture Layout_First Floor

From Vangviet.com

Survey

An often underestimated form of data collection is the employee interview. This is the opportunity to question those people who may have seen the pest and can describe things about the encounter that wouldn't fit on the pest sighting log. This isn't meant to be an interrogation but a fact finding and discussion opportunity. When looking for someone to discuss seeing pests, most people will go to the manager; this usually will result in limited info or second hand info at best.

The best people in the facility to question are the ones who are actually involved in the jobs which place them in contact with the pests, i.e. bus boys, dishwashers, janitors, etc. The key questions you need to ask are:
Who saw the pest?
Followed by
What did you see?
Where did you see it?
When did you see it?
Where did it go?
The importance of asking open ended questions can't be over stated since this will get the client talking and may elicit information no one thought to ask. Be sure to document all the particulars in case you need to follow up.

Interpretation of the data

The interpretation of the collected data is what separates the best companies from the rest. Taking all the varied pieces of data and analyzing them is analogous to a doctor providing a diagnosis, a detective piecing clues together or a stock trader deciding which company to invest in and all at the same time. Knowledge and experience go a long way to helping with the interpretation, but it's a process that can be learned easily and quickly.
What do the pests need to survive? Food, Water, Shelter and they prefer areas with little air movement. During your data collection where did you find these areas?

Food
Were these long term problems or singular occurrences?
What about seasonal variation?
Do they cook on the premises?
Is the food bagged?
Sanitation issues?
Trash collection (Do they use self closing garbage cans?)
How, when? Is there organic debris build up?

Water
Continual moisture issues?
A/C issues?
Landscaping and substrate material?
Ponds, fountains lakes?

Shelter
Storage areas, how cluttered?
Product rotation, first in and first out?
Housekeeping issues?
Maintenance issues?
Construction issues?
Lighting outside; is it on the building or shining onto it?

Monitors
Where are the pests?
How are they distributed on the glueboards?
What gender is on the glueboard? What life cycle stage is present?
You have gathered all this information as part of a primary survey during your data collection; you have documented any discrepancies or concerns and now have a clear picture of potential problems.
The interpretation of collected data as you see is a complex process, but as we said one that can be learned. The key to all good interpretation is to Keep It Simple.
What info can help you break the triangle?

Following this inspection you should have a good clear understanding of the customers needs and what you can do to solve their problems.

Environmental Adjustment

At this stage you take the information about the structure, combine it with the pest signs and your knowledge about the biology of the pests to complete your treatment recipe. Treatment strategies depend a great deal on the company's philosophy of material use. The tide of public opinion has come full circle to a more natural control method and strategy. This doesn't mean to eliminate pesticides, but it does mean to justify and document the reasons for their use, and a judicious approach to application.

Consideration must be given first to the reduction of any pests present, second to eliminating the potential for more pests to enter and lastly correct facility issues. These facility issues may entail not just structural modifications but cultural modifications via education to help in the pest elimination process.

Getting rid of the pest may involve just vacuuming or placing out snap traps and glueboards. Think first of non chemical means of getting rid of the pest, especially if they are adults.

Eliminating the pest entrance and correcting facility issues will fall under the heading of habitat modification. A few examples of this are screens on windows, sealing cracks and openings, and placing copper mesh in weep holes. Some questions to ask regarding the facility are:

1. What activities can the client do to reduce pest attraction to the building or space?

2. What modifications can be made to the structure to eliminate pest entrance?

3. What surrounding areas can be modified to keep pests from building up in large numbers?

The environmental adjustment of the structure and its surrounding area is simply anticipating the pest's activity and taking measures to stop it. Eliminating attraction to a space or structure is easier and more effective than eliminating the source of pests. The attraction reduction anticipates the pest pressure and source reduction takes care of the pest after it has entered.

Sanitation is a key component in environmental adjustment. It is the key to the control of flies, roaches, ants and rodents, just to name a few. Clients often fail to adhere to our recommendations or to even clean, expecting you as the pest management professional to "just treat". An effective method to reduce the sanitation issue is to use biological materials to enhance the sanitation process, reducing the need for arbitrary and often unnecessary treatment with products.

Customers see the thoroughness of the application and feel the value of your service. Products increase in their efficacy when no competing food sources are available. This treatment can be applied to cove areas, trash receptacles, drains, equipment bases, walls and any areas that won't contact food.

The exterior reduction of attraction and environmental adjustment involves sanitation of the trash receptacles including dumpsters, changing lighting from mercury vapor to sodium vapor, brushing webs from overhangs and areas around entrances, and placing rodent stations with snap traps in them next to entry doors, keeping vegetation 18" from the building to name a few.

Now gather all the documentation, support information and the proposal and go sell the account. It may seem like a lot to go through, but you'll see the sales come in more readily when you have a formal plan, when you have demonstrated your abilities by inspecting the building and when you produce a formal proposal. (See Appendix F)

Proposal

It is becoming increasingly common to have a proposal faxed or emailed to the client. If there are no distance issues or other valid reasons for not visiting the customer, you should always go to see the customer. You want to be there to demonstrate your sincere desire to be part of the process and a strategic partner with the new customer. If you send the info and ask for the signature, the customer immediately has the impression they are a number. Remember they are a partner now and a great source of networking opportunities for you. Get their business card or have other clients of your available if they have a need; if you service a dry cleaner then let them know you recommend them (have them mention this to the dry cleaner when they go). Whatever need they may have; you probably know someone you can recommend. That is being a strategic partner.

When delivering the proposal and discussing your potential strategic alignment, here are some questions to ask the potential buyer, from author Tom Searcy:

1. In terms of time, money, and risk, what business problem will working with us solve for you? This answer will tell you whether the prospect has a real issue they are willing to pay to resolve or if they're just curious. Companies tire-kick in the market for alternative vendors all the time. Sometimes their rules require them to do this, other times they are keeping their incumbent vendor "honest" on price and features and there is always the desire to stay current. Regardless, if there is not a sizable business problem that they are looking to solve, your prospect is just curious, not really interested. No interest, no engagement, no money.

2. How will you measure our success together 60 days after we start work? Success measures, specific measures are a great determinant of sincerity and seriousness in the consideration process of a buyer. The reason that we limit our review period to 60 days is because we want to determine urgency and clarity. If you get urgency and clarity at the right level, you are talking to someone who is going to make a decision faster.

3. How much better does our "better" have to be for you to work with us? I love threshold questions. Thresholds set a defined bar as to how good something has to be for there to be action taken. You can first find out if there is such a bar determined to assess sincerity of interest, then you can decide if you can make that bar. Both answers can keep you from wasting your time.

4. How soon does this problem need to show improvement for you to feel that our work together is successful? Another type of threshold question, this focuses on urgency and one of the leverage points that smaller companies have- speed. If your client can see success way out in the future as being their goal, chances are that anyone can be their partner. When that is the case, you have lost an important advantage and your bigger branded competitors are the more likely winners.

5. What process will you follow in bringing us on as your provider? Process questions let you know how far out your prospect has thought through bringing on a replacement solution provider. If they have no answer to this question, then you are probably talking to the wrong level of person, or they are not seriously considering a big change.

Getting a higher conversion rate is about pitching fewer losers so you have more time for the winners- pick the high potential winners earlier by not leaving it to chance. Ask the hard questions and let the hard facts guide your path.

Author, speaker and consultant Tom Searcy is the foremost expert in large account sales. With Hunt Big Sales, he's helped clients land more than $5 billion in new sales.

The Last Word
At the start of this manual we made some assumptions, such as you already have basic pest management experience and knowledge. Innovative pest management is something that any technician can do and any sales person can sell, but like anything you want to do well, it needs to be practiced. The steps may seem overwhelming, but with the practice it gets easier and with experience can become second nature, as in this is our philosophy of treatment. This is the basis for the sustainable or Green pest management everyone wants to offer, but may not be able to. My hope in writing this is to educate you and give some opinions and observations based on 30 years of training and working with technicians and sales people in the field. Thank you for your time.

About the author:

Jerry Hatch owner of Hatch LLC has spoken at numerous Pest Control Associations, Environmental Health associations, Service Companies, Public Health Departments and Hospitality industry meetings, and was the guest lecturer at Eastern Michigan University Hospitality degree program. Over his career he has written numerous articles for trade associations and local newspapers, including Holmes on Holmes magazine and been a guest on several radio programs including the Dave FM morning show.
He can be reached at: bce9083@gmail.com

Appendix A
Integrative Pest Management
Data Collection
- Inspect the areas of concern
    - Look for signs of
        - pests
        - food
        - moisture
        - potential entrance areas
        - Harborage areas
    - Check monitors (for pests or parts)
        - Small monitors
        - ILT system
    - Survey the customer and employees

Interpretation
- Identify conditions conducive to pest harborage
    - Attractants
    - Organic debris build up
- Identify terminal points of infestation or pest movement
- Identify non chemical measures

Environmental Adjustment
- Anticipate pest activity (void or entrance area treatment and exclusion)
- Sanitation material application via sprayer (compressed air or trigger applicator) to all areas located as:
    - Organic debris build up
    - Cove base areas

- o Equipment bases and contact points with adjacent walls, machines, etc.
- o Drains
- Treat the trash areas
    - o Dumpster area
    - o Trash cans
    - o Rubber matting
- Treat organic debris build up with Biological materials
    - o Drains
    - o Equipment and floor contact areas
    - o Threshold areas
- Chemical Treatment where necessary
    - o Pest Harborage
    - o Potential areas of entrance
    - o Pest shelters

Communication & Documentation

- Make recommendations
- Complete documentation
- Review with the customer

Notes:

Appendix B
Pests

Location	Person Reporting	What was seen	Treatment date	Result

Appendix C

Technician _____   **Trainer** _____

Verifiable IPM Training for Residential Services

Integrated pest Management forms the basis for all treatment regimens especially in Green accounts. We believe it's the foundation of all our service training. The following are the practical or field competencies all of our technicians have been trained in.

	Initials	
IPM	Tech	Trainer
Definition and overall philosophy of IPM or inspection based services	____	____

Inspection and Harborage Identification

Technician has all the tools for inspection   ____   ____
Flashlight
Clipboard
Probing Tool
Extended Mirror

Technician demonstrates the ability to locate possible moisture sources   ____   ____
such as
Pipe accesses or plumbing penetrations
Drain areas
Threshold covers
Aquariums
Pet bowls
Sauna/Jacuzzi
Others

Technician demonstrates ability to locate organic debris sources   ____   ____
Mulch areas
Threshold plates
Edges of Decking
Door frames
Baseboard areas
Gutters

Others

Technician demonstrates ability to find various harborage or pest entry _____ _____
points such as
Doorways
Windows
Pipe or plumbing penetrations
Weep holes
Foundation gaps or cracks in walls
Other

Technician demonstrates ability to inspect equipment conducive _____ _____
to pest presence and harborage such as
Deck furniture
BBQ Grill
Planters
Pool house
Children's Playscapes
Sheds
Others

Identification

Technician demonstrates ability to identify main pest problems including _____ _____
adults, immature stages and damage from
Rodents
Ants
Roaches
Stinging insects
Stored Product Pests
Occasional Invaders
Fleas, Ticks, Bed bugs
Others

Management Decision

Technician demonstrates ability to choose products based on their _____ _____
Chemical make up
Formulation
Application availability
Environmental Quality (Green)
Others

Application

Technician demonstrates the ability to treat various areas with approved _____ ____
techniques and according to label requirements
_____ Void
_____ Crack and crevice
_____ Spot Application
_____ Broadcast
_____ Bait gel placement
_____ Web brush
_____ Perimeter broadcast
_____ Space spray
_____ ULV
_____ Others _____

Technician demonstrates knowledge of maintenance and use of _____ _____
various equipment such as
Compressed air sprayer
Accu sprayer
Getz duster
Aerosol dispersal equipment, system 3
Bait gun
Spreaders
Others _____

Documentation

Technician demonstrates the ability to document all the following _____ ____
Conditions conducive to pests
Entry areas
Customer corrections
Product use
Precautionary statements
Others

Technician demonstrates the use of communication tools _____ ____
Summary discussion with the client
Others

Appendix D

Technician Name_____   **Trainer**_____

Verifiable IPM Training for Commercial Services

Integrated pest Management forms the basis for all treatment regimens whether it's an AIB, HACCP or Green account. We believe it's the foundation of all our service training. The following are the practical or field competencies all of our technicians have been trained in.

                                         Initials

IPM   Tech Trainer

Definition and overall philosophy of IPM or inspection based services ____ ____

Inspection and Harborage Identification

Technician has all tools for inspection   ____ ____
Flashlight
Clipboard
Probing Tool
Extended Mirror
Others

Technician demonstrates the ability to locate possible moisture sources ____ ____
such as
Pipe accesses or plumbing penetrations
Drain areas
Threshold covers
Evaporation pans
Steam machines or equipment
Others

Technician demonstrates ability to locate organic debris sources   ____ ____
Drain areas
Threshold plates
Edges of equipment
Door frames
Baseboard areas
Others

Technician demonstrates ability to find various harborage or pest entry _____ _____
points such as
Doorways
Windows
Pipe or plumbing penetrations
Other

Technician demonstrates the ability to place and read various monitors _____ _____
Minimum placement in a food service facility food prep area is 15 monitors
Trapper rodent monitors
Trapper Trifold monitors
Others

Technician demonstrates ability to inspect food equipment conducive _____ _____
to pest presence and harborage such as
Food preparation machines
Meat Slicers
Kettles
Stoves
Cleaning equipment
Ice machines
Others

Identification

Technician demonstrates ability to identify main pest problems including _____ _____
adults, immature stages and damage from
Rodents
Ants
Roaches
Flies
Stored Product Pests
Occasional Invaders
Fleas, Ticks, Bed bugs
Others

Management Decision

Technician demonstrates ability to choose products based on their _____ _____

Chemical make up
Formulation
Application availability
Environmental Quality (Green)
Others

Technician demonstrates ability to mix and use bio remediation materials _____ _____
As part of all food service treatment specifications it is understood that sanitation plays an important role. **We apply bio remediation products to all food service locations as part of an effort to increase the level of sanitation and reduce the need for pesticides.**

Application

Technician demonstrates the ability to treat various areas with approved _____ _____
techniques and according to label requirements
_____ Void
_____ Crack and crevice
_____ Spot Application
_____ Broadcast
_____ Bait gel placement
_____ Web brush
_____ Perimeter broadcast
_____ Space spray
_____ ULV
_____ Others _____

Technician demonstrates knowledge of maintenance and use of _____ _____
various equipment such as
Compressed air sprayer
Accu sprayer
Getz duster
Aerosol dispersal equipment, system 3
Bait gun
Spreaders
Others _____

Documentation

Technician demonstrates the ability to document all the following _____ _____
Conditions conducive to pests
Entry areas

Customer corrections
Product use
Precautionary statements
Others

Technician demonstrates the use of communication tools _____ ____
Logbook
Summary discussion with the client
Urgent needs form
Others

Appendix E
© A & C Black Publishers Ltd 2006

Analyzing Your Business's Strengths, Weaknesses, Opportunities, and Threats

GETTING STARTED

SWOT analysis (Strengths, Weaknesses, Opportunities, and Threats) is a method of assessing a business, its resources, and its environment. Doing an analysis of this type is a
good way to better understand a business and its markets, and can also show potential investors that all options open to, or affecting a business at a given time have been thought about thoroughly.

The essence of the SWOT analysis is to discover what you do well; how you could improve; whether you are making the most of the opportunities around you; and whether there are any changes in your market—such as technological developments, mergers of
businesses, or unreliability of suppliers—that may require corresponding changes in your business. This action list will introduce you to the ideas behind the SWOT analysis, and
give suggestions as to how you might carry out one of your own.

FAQS

What is the SWOT process?

The SWOT process focuses on the internal strengths and weaknesses of you, your staff, your products, and your business. At the same time, it looks at the external opportunities
and threats that may have an impact on your business, such as market and consumer trends, changes in technology, legislation, and financial issues.

What is the best way to complete the analysis?
The traditional approach to completing SWOT is to produce a blank grid of four columns— one each for strengths, weaknesses, opportunities, and weaknesses—and then
list relevant factors beneath the appropriate heading. Don't worry if some factors appear in more than one box and remember that a factor that appears to be a threat could also represent a potential opportunity. A rush of competitors into your area could easily represent a major threat to your business. However, competitors could boost customer numbers in your area, some of whom may well visit your business.

What is the point of completing a SWOT analysis?
Completing a SWOT analysis will enable you to pinpoint your core activities and identify what you do well, and why. It will also point you towards where your greatest opportunities lie, and highlight areas where changes need to be made to make the most of
your business.

MAKING IT HAPPEN
Know Your Strengths
Take some time to consider what you believe are the strengths of your business. These could be seen in terms of your staff, products, customer loyalty, processes, or location.
Evaluate what your business does well; it could be your marketing expertise, your environmentally-friendly packaging, or your excellent customer service. It's important to try to evaluate your strengths in terms of how they compare to those of your competitors.
For example, if you and your competitors provide the same prompt delivery time, then this cannot be listed as a strength. However, if your delivery staff is extremely polite and helpful, and your competitor's staff has very few customer-friendly attributes, then you should consider listing your delivery staff's attitude as a strength. It is very important to be totally honest and realistic. Try to include some personal strengths and characteristics of your staff as individuals, and the management team as individuals. Whatever you do,
you must be totally honest and realistic: there's no point creating a useless work of fiction!

Recognize Your Weaknesses
Try to take an objective look at every aspect of your business. Ask yourself whether your products and services could be improved. Think about how reliable

your customer service is, or whether your supplier always delivers exactly what you want, when you
want it. Try to identify any area of expertise that is lacking in the business. as you can then take steps to improve that aspect. For example, you might realize that you need some more sales staff, or financial help and guidance. Don't forget to think about your
business's location and whether it really does suit your purpose. Is there enough parking, or enough opportunities to attract passing trade? Your main objective during this exercise is to be as honest as you can in listing weaknesses. Don't just make a list of mistakes that have been made, such as an occasion when a customer was not called back promptly. Try to see the broader picture instead and
learn from what happened. It may be that your systems or processes could be improved
so that customers are contacted at the right time, so work on boosting your systems and making that change happen rather than looking about for someone to blame. It's a good idea to get an outside viewpoint on what your weaknesses are as your own
perceptions may not always marry up to reality. *You* may strongly believe that your years of experience in a sector reflect your business's thorough grounding and knowledge of all
of your customers' needs. Your customers, on the other hand, may perceive this wealth of experience as an old-fashioned approach that shows an unwillingness to change and work
with new ideas. Be prepared to hear things you may not like, but which, ultimately, may be extremely helpful.

Spot the Opportunities
The next step is to analyze your opportunities, and this can be tackled in several ways. External opportunities can include the misfortune of competitors who are not performing
well, providing you with the opportunity to do better. There may be technological developments that you could benefit from, such as broadband arriving in your area, or a new process enhancing your products. There may be some legislative changes affecting
your customers, offering you an opportunity to provide advice, support, or added services. Changes in market trends and consumer buying habits may provide the development of a niche market, of which you could take advantage before your competitors, if you are quick enough to take action. Another good idea is to consider your weaknesses more carefully, and work out ways of

addressing the problems, turning them around in order to create an opportunity. For example, the pressing issue of a supplier who continually lets you down could be turned into an opportunity by sourcing another supplier who is more reliable and who may even
offer you a better deal. If a member of staff leaves, you have an opportunity to reevaluate duties more efficiently or to recruit a new member of staff who brings additional experience and skills with them.

Watch Out for Threats
Analyzing the threats to your business requires some guesswork, and this is where your analysis can be overly subjective. Some threats are tangible, such as a new competitor moving into your area, but others may be only intuitive guesses that result in nothing.
Having said that, it's much better to be vigilant because if potential threat does become a real one, you'll be able to react much quicker: you'll have considered your options already and hopefully also put some contingency planning into place.
Think about the worst things that could realistically happen, such as losing your customers to your major competitor, or the development of a new product far superior to your own. Listing your threats in your SWOT analysis will provide ways for you to plan to deal with the threats, if they ever actually start to affect your business.

Use Your Analysis
After completing your SWOT analysis, it's vital that you learn from the information you have gathered. You should now plan to build on your strengths, using them to their full potential, and also plan to reduce your weaknesses, either by minimizing the risk they
represent, or making changes to overcome them. Now that you understand where your opportunities lie, make the most of them and aim to capitalize on every opportunity in front of you. Try to turn threats into opportunities. Try to be proactive, and put plans into
place to counter any threats as they arise. To help you in planning ahead, you could combine some of the areas you have
highlighted in the boxes; for example, if you see an external opportunity of a new market growing, you will be able to check whether your internal strengths will be able to make
the most of the opportunity. For example, do you have enough trained staff in place, and can your phone system cope with extra customer orders? If you have a weakness that undermines an opportunity, it provides a good insight as to how

you might develop your internal strengths and weaknesses to maximize your opportunities and minimize your threats. The basic SWOT process is to fill in the four boxes, but the real benefit is to take an
overview of everything in each box, in relation to all the other boxes. This comparative analysis will then provide an evaluation that links external and internal forces to help
your business prosper.

COMMON MISTAKES
Focusing just on a few issues
Don't just focus on the large, obvious issues, such as a major competitor encroaching on your business. You need to consider all issues carefully, such as whether your Internet
system provides everything you need or whether your staffing levels are as they should be.
Completing your SWOT analysis on your own
Do take advantage of other people's contribution when you're completing your SWOT analysis; don't try and do it alone. Other people's perspectives can be very useful,
particularly as they may not be as close to the business as you are. This distance can often help them see answers to thorny questions more easily, or to be more innovative: we all get stuck in a rut at points.
Using your analysis for the next ten years
Don't do a SWOT analysis once and then never repeat the exercise. Your business environment will be constantly changing, so use SWOT as an ongoing business analysis
practice.
Relying on SWOT to provide all the answers
Use SWOT as part of an *overall* strategy to analyze your business and its potential. It is a useful guide, not a major decision-making tool so don't base major decisions on this
analysis and nothing else.

Appendix F
15 Deal Breakers to Avoid When Pitching an Idea
From the Art of Manliness at artofmanliness.com

You've finally gotten a meeting with the people who can turn your dream into a reality. You can't wait to walk into that room and sell them your idea.

Awesome. But here's one of the most important things you need to know: The buyer is not looking to say yes. *They're looking to say no.*

This is hard for the seller to understand. You feel like the buyer is just waiting to hear your world-changing idea. You're one guy, with one idea, and you've been working on that idea for years. It's all you think about.

But the buyer sees dozens, hundreds, even thousands of guys just like you every year. You're a dime a dozen. For them, saying no is the easiest option. Saying yes involves risk—of their money and reputation–and it involves time, hassle, and responsibility. Saying no simplifies their life and lets them get on with their day. Basically, buyers are looking for any reason to turn you down.

Because of the number of pitches they get, all buyers develop ways of slotting sellers into yes and no categories. Your train can be chugging right along, but if you raise a deal breaker red flag—they'll throw the switch and put you on the no track. These flags can be really small things, but they've probably found that 8 out of 10 people who exhibit those traits end up being a nightmare to work with. And they're not willing to gamble that you're one of the two who are exceptions to the rule.

Sure, buyers' deal breakers aren't fair—not at all. Your idea might be truly fantastic, but you're having a terrible day and thus blow the pitch. But buyers can't give every pitch the same attention and thus develop a sorting system by necessity.

Even though buyers' deal breakers aren't fair, they are happily pretty easy to avoid. Here are 15 pitching pitfalls to avoid stepping into, as gleaned from Stephanie Palmer's *Good in Room* (as an executive at MGM, she ruined many a screenwriter's day) and my personal experience on both sides of the desk.

1. Arriving late. Showing up late demonstrates that you don't respect the buyers' time. Here's a good maxim to live by: "If you're on time, you're late." There are always going to be unexpected obstacles to getting into that meeting room—there's surprisingly heavy traffic on the way there, you have to park a few blocks away, you have to go through a security check in the lobby, the office is on the

50th floor, and all the elevators are full. So you should plan on pulling into the general vicinity of the meeting place 15 minutes ahead of time. If you don't encounter any of the obstacles just mentioned, then when you get to the office early, tell the receptionist you're there, but that there's no need to announce you until 5 minutes prior to the meeting time. Then just take a seat in the waiting area and review your notes.

2. Dressing inappropriately. Dress in line with the standard of the company you're pitching to. If they're a traditional, conservative business, wear a suit. If they're a modern and casual business, wear khakis and a sport coat. Consider wearing something blue as this color engenders a feeling of trust.

3. Taking the wrong seat. People are strangely territorial about their seats. Just try sitting in the wrong pew at a small church (families actually used to "rent" a pew back in the 18th century for the privilege of having their name emblazoned on it).

Sit in the wrong seat at a pitch meeting, and someone may have to awkwardly say, "That's my seat." Or they may say nothing, but sit through the meeting feeling a bit put out by your perceived presumptuousness.

Where they'd like you to plant your kiester may be obvious—but if it's not, then simply ask, "Where would you like me to sit?" when you walk in.

4. Getting their name wrong. Everyone loves the sound of their own name, which is why using someone's name is one of the easiest ways to build rapport. Conversely, getting someone's name *wrong* is one of the quickest ways to stop rapport-building dead in its tracks.

This might seem like a no-brainer, but I can't tell you how many emails we get addressed to "*Brent* and *Kay*."

When you get someone's name wrong, you show you really don't know much about the company you're pitching to or that you're inattentive to details. It can also make you seem highly disingenuous if you follow your name-blunder with, "I'm such a big fan of yours!"

5. Not addressing the pitch to everyone in the room. If both the president and the VP are sitting in on the meeting, don't only address your remarks to the president, and slight the veep. Talk and make eye contact with everyone in the room, from the lowliest note taker to the head honcho.

6. Acting nervous. Maybe your idea is great, you prepared for the presentation like a champ, and the nervousness you're exhibiting is simply from a fear of public speaking. But there's no way around it: nervousness translates as incompetence and weakness. The buyer will wonder if you didn't prep enough or if your idea is so risky that even you don't have full confidence in it. Either way, you've just made your job ten times harder. And you've made their job more difficult as well; they might like your idea, but feel like they can't introduce you to the higher-ups.

Nervousness can be manifested through fumbling with materials, technical glitches, excessive "ummms" and "uhhhs," and super sweaty pits. If the latter is a problem for you, wear a jacket and/or wear a clinical strength deodorant.

7. Starting with an apology. Whether for your lateness, your nervousness, or something, else, this is quite possibly the weakest opening you can give your pitch. Let the first words out of your mouth be a show of strength and confidence.

8. Giving your own opinion of your work. Don't say, "This is an awesome idea that is going to change the world." Let the idea speak for itself.

9. Telling the buyer how they're going to feel. Don't say, "You're going to love this" or "I have an idea that's perfect for you." People hate being told what they think or how they're going to feel.

10. Jumping into your pitch too soon. The first thing you want to do is build rapport with the buyer. Jumping into your pitch before you build that rapport is like trying to dive down a Slip 'n Slide before you've turned on the water.

11. Talking money too soon. If you're looking for a big investment, and you talk about that nut too soon, the buyer is going to feel immediate trepidation and view the rest of your presentation through the lens of, "This better be good to warrant that amount of money!" It heightens their expectations considerably. But if you dazzle them with your presentation, by the time you get to talking money, they'll see the number through the lens of, "Whatever it is, we'll make it work. We have to make this happen."

12. Offering phony flattery. A company recently made me a pitch. They began their Powerpoint presentation with a slide that said, "The Art of Manliness: World's Best Online Magazine for Men." A spreadsheet they sent us was entitled: "Art of Manliness World's Best Data." Did I mention that the meeting reminder they sent called the meeting "Art of Manliness+World's Best" and the password was "TheBest?"

To me this came off as desperate and over-the-top. A little flattery is good and builds rapport. But too much comes off as insincere and desperate—as it will make the buyer feel like what you're selling needs to be unduly padded.

To flatter with class, compliment the buyer on something specific they've done that you liked, especially something that the average joe who doesn't know a lot about the company wouldn't be aware of.

13. Not giving enough context. In the book *Made to Stick*, Chip and Dan Heath discuss what they call "The Curse of Knowledge." The Curse of Knowledge describes the fact that when you're steeped in a subject, you can easily forget that others are not as familiar with it as you are. Something may seem so basic to you that it doesn't even warrant mentioning, but for someone else, it can be a brand new idea. By assuming that the buyers know things that they don't, you may omit key facts from your presentation. The buyers' resulting confusion will then lead to writing you off.

If there are spots in your pitch where you're not sure if you and the buyer are on the same page, simply say, "Are you familiar with X?" before launching into your next point. This also keeps you from boring the buyer with information they already know.

14. Using terminology the buyer isn't familiar with. This is related to the point above. We had a television/film agent who would talk to us with lots of Hollywood lingo that a couple of Oklahomans could not follow. And that's part of the reason we switched to another agent.

15. Saying just "I don't know." Instead say, "I don't know. But I will find that out for you and send you an email with the information later today."

www.ingramcontent.com/pod-product-compliance
Lightning Source LLC
Chambersburg PA
CBHW081904170526
45167CB00007B/3143